Industrial radiography

Ionising Radiation Protection Series No 1

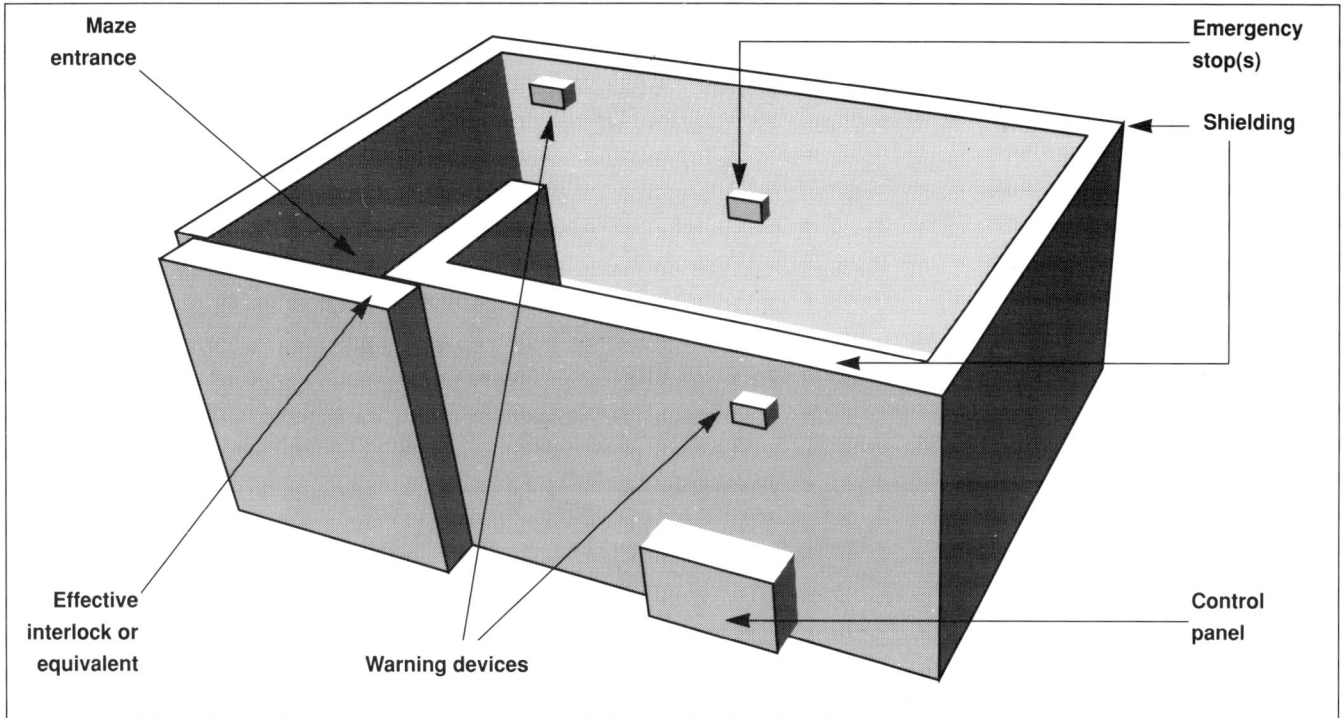

Schematic enclosure for 'routine' radiography

Introduction

This information sheet is for directors and managers of industrial radiography companies (and their clients). It summarises precautions which, if followed, should help to ensure compliance with the main requirements of the Ionising Radiations Regulations 1985 (IRR85), especially the requirement to ensure that exposures are as low as reasonably practicable. Alternative approaches which satisfy IRR85 are possible.

Inadequate control of industrial radiography can lead to substantial exposure of radiographers and others to radiation. A number of serious incidents have occurred because of failure to maintain equipment, to carry out routine monitoring and to employ proper emergency procedures. In 1991 more industrial radiographers than nuclear industry workers had a high annual dose (greater than 15 millisieverts).

Routine radiography

In most premises where readily moveable articles are regularly radiographed, it should be reasonably practicable for the occupier (or a contractor) to provide a permanent shielded enclosure for such work. Cranes or fork-lift trucks may be needed to transfer larger articles to the enclosure. Normally, routine radiography should not be carried out in open workshops (IRR 85, reg 6(2)).

The design and construction of the walls of the enclosure should be sufficient to encompass the whole of the controlled area (IRR85, reg 8), ie the dose rate outside should be below 7.5 microsieverts (μSv) per hour. HSE inspectors have found badly designed enclosures in light industrial units where dose rates have exceeded 7.5 μSv per hour next door! People working near or above the enclosure in cranes or office/storage areas may also be at risk. Scattered radiation (sky-shine) outside open-top enclosures can often be controlled by adequate collimation. If you decide to use a different source or generator you may have to upgrade the shielding. Your Radiation Protection Adviser (RPA) should be consulted.

For X-ray generators it should be reasonably practicable to install effective devices, eg reliable electrical or mechanical interlocks, which prevent or terminate an exposure if the door of the enclosure is opened. It may not always be possible to achieve this level of protection for sealed sources. These devices should be installed so they are fail-safe, but they can deteriorate and require periodic checks. Clear warning signals will also be needed (IRR85, reg 6(2)).

Site radiography

'Site radiography' is only acceptable when it is not reasonably practicable to provide a shielded enclosure for an article to be radiographed (IRR85, reg 6(2)).

Contractual arrangements with clients should allow adequate time to plan for: localised shielding, barriers, warning notices and signals, lighting, location of the control point etc. HSE will normally require seven days prior notice of the work (IRR85, reg 5(4)). Careful planning is vital for complex work, eg work in structures where access to the 'controlled area' is difficult or can be made at different levels or where the person changing films is not clearly visible from the control position. A fail-safe communication system between the radiographer and assistants will prevent misunderstandings.

Normally you require barriers and warning signals to demarcate the 'controlled area' within which radiography takes place (IRR85, reg 8(5)). Also, you should make reasonably practicable arrangements, with the client if necessary, to provide localised shielding, eg appropriate collimation, moveable panels, pre-formed shields, lead mats, or bags of lead shot, to restrict the size of the controlled area (IRR85, reg 6(2)).

Failure of critical components of radiography equipment such as the control cable, pigtail connector and guide tube of projection type containers is likely to leave the source exposed. So it is essential you have a programme of inspection and maintenance for this equipment (IRR85 regs 6, 25). Source changing should be left to specialist contractors unless you have the correct equipment and trained staff capable of carrying out this work safely.

General

Dose rate monitoring instruments enable radiographers to check that sealed sources have fully retracted into containers, or X-ray sets have stopped emitting radiation. Periodic monitoring outside an enclosure is needed to check that shielding remains adequate (IRR85 regs 6,8,24,25). The monitoring equipment must be calibrated by a 'qualified person' and maintained in good working order (IRR85, reg 24). Many radiographers have received inadvertent exposures to radiation because monitoring equipment was either defective or unavailable.

Your written local rules outlining the systems of work required for radiography should be available to employees (IRR85, reg 11). Supervisors should be made aware of any special arrangements for specific sites (IRR85, regs 6,12). You should prepare a contingency plan to deal with reasonably foreseeable 'emergencies' such as a stuck source (IRR85, reg 27).

A typical 'emergency' kit to be taken to the site would normally include bags of lead shot, a shielded pot and long-handled tools. A plan will only be effective if it is rehearsed periodically, eg using dummy sources.

The RPA you appoint needs up-to-date knowledge in the requirements of IRR85 and radiation protection for industrial radiography, as well as enough time (IRR85, reg 10). The RPA will often be appointed from outside the company. Radiation Protection Supervisors are needed to ensure the work is carried out in accordance with the local rules (IRR85, reg 11).

You rely on trained radiographers to follow safe systems of work in order to restrict exposure so far as reasonably practicable. Regular refresher training helps radiographers recognise the importance you attach to the restriction of exposure.

Your radiographers will probably need to be designated as 'classified persons' (IRR85, reg 9). If so, you should arrange for an approved dosimetry service to make routine dose assessments and keep dose records (IRR85, reg 13). It is sensible to provide alarming dose-rate meters as well. You should ensure that there are clear instructions for issuing and wearing dosemeters and secure facilities for their storage when not being worn. Dosemeters should not be left in toolboxes and overalls or inside radiography enclosures, where they may be damaged or exposed when not being worn.

Further information

HSC *The protection of persons against ionising radiation arising from any work activity: The Ionising Radiations Regulations 1985* Approved Code of Practice Parts 1 and 2 1985 ISBN 0 7176 0508 6

HSC *Dose limitation - restriction of exposure: additional guidance on regulation 6 of the Ionising Radiations Regulations 1985* Approved Code of Practice Part 4 1991 ISBN 011 885605 7

Oil and Chemical Plant Constructors' Association *Radiation safety for site radiography* Kluwer Publishing 1986 ISBN 0 903393 86 7

This publication may be freely reproduced, except for advertising, endorsement or sale purposes. The information it contains is current at 2/94. Please acknowledge the source as HSE.

HSE
Health & Safety
Executive

A FRAMEWORK

for the restriction of

occupational exposure

to

IONISING

RADIATION

HS(G) 91

London: HMSO

HS(G) series

The purpose of this series is to provide guidance for those who have duties under the Health and Safety at Work etc Act 1974 and other relevant legislation. It gives guidance on the practical application of legislation, but it should not be regarded as an authoritative interpretation of the law.

Enquiries regarding this or any other Health and Safety Executive publications should be made to:

HSE Information Centre
Broad Lane
Sheffield S3 7HQ
Telephone: 0742 892345
Fax: 0742 892333

HS(G) 91

ISBN 0 11 886324 X

CONTENTS

INTRODUCTION

1   This guidance gives advice on the duty under Regulation 6 of the Ionising Radiations Regulations 1985 (IRR 85)[1] to restrict so far as reasonably practicable the extent to which persons at work are exposed to ionising radiations (the ALARP requirement). The publication is relevant to all work with ionising radiation ranging from the use of articles containing small amounts of radioactive substances, for example industrial gauges, to complex nuclear plant; but it is particularly concerned with activities which may involve significant exposure to ionising radiations. It does not give advice on the restriction of public or medical exposure.

2   The publication is aimed at managers who have responsibilities for work with ionising radiations and at employee representatives. It gives advice about the type of action required to ensure that an effective and consistent approach is taken to the restriction of occupational exposure. The guidance outlines the importance of an explicit commitment by senior management to the aim of restricting occupational exposure as far as reasonably practicable, the need for suitable arrangements to implement this policy, and the value of reviewing the work periodically to ensure that exposures continue to be adequately controlled. However, it does not attempt to give a detailed explanation about the measures that may be necessary to comply with Regulation 6 IRR85.

3   It is recognised, however, that the application of sound radiation protection principles by competent personnel will often be all that is necessary for many of the everyday decisions that have to be taken about the restriction of exposure to ionising radiation.

4   The guidance is concerned with normal work practices where occupational exposure to ionising radiation is almost certain to occur at some time. It does not specifically address arrangements for reducing the probability of accidental events occurring which might have serious radiological consequences. However, these events will need to be considered in accordance with Regulation 25 IRR 85.

COMMITMENT OF MANAGEMENT

5   It is essential that senior management demonstrates an active and continued commitment to the aim of restricting exposure to ionising radiation. This commitment should be clearly stated in the written policy for health and safety required by the Health and Safety at Work etc Act 1974 (HSWA) and reflected in the day-to-day practices of senior managers and directors.

ORGANISATION FOR RESTRICTING EXPOSURE

General

6   To translate this commitment into effective action, senior management should identify appropriate organisational arrangements, assign clear responsibilities to put these into effect and seek to establish a culture within which all those in the organisation recognise the importance of restricting exposure to ionising radiation.

Radiation protection programme

7   Every employer who undertakes work with ionising radiation will require an overall radiation protection programme to meet the specific requirements of IRR 85, for example: to designate classified persons and to assess their doses; to designate and monitor controlled areas; to prepare (and update) 'local rules'; to appoint a radiation protection adviser (RPA) and one or more radiation protection supervisors (RPS); and to provide training. However, the ALARP requirement underpins the whole of IRR 85; therefore the arrangements provided to meet this requirement should normally be an integral part of that programme.

8   The detailed provisions of the programme will depend on the scope of the work with ionising radiation and the magnitude of any exposure likely to arise from the work (and hence the relevance of specific requirements under IRR 85). The Approved Code of Practice[2] (ACOP) supporting IRR 85 provides guidance on certain elements of the programme, including, in particular, the key role of the RPA.

9   Appropriate senior managers should be given overall responsibility for ensuring that the programme is:

(i)   fully implemented; and

(ii)   reviewed from time to time to evaluate its effectiveness and to identify inadequacies in the arrangements for restricting exposure.

Responsibilities of managers and role of advisers

10   The written health and safety policy, organisational diagram, job descriptions or other documents should clearly define the responsibilities of managers in relation to restriction of exposure to ionising radiation. Radiation protection supervisors appointed under Regulation 11 IRR85 should have a special supervisory role for ensuring either directly or indirectly that the work is done in accordance with 'local rules'. It is important that those managers who have been assigned responsibilities for radiation protection understand those responsibilities and are in a position to discharge them.

11   Senior management should also ensure that the functions of the radiation protection adviser (including relevant matters listed in paragraph 75 of the ACOP Parts 1 and 2[2]), supporting health physicists, the medical adviser and any other specialist who provides advice to local managers are clearly defined. Line managers should be made aware of the roles of these individuals and be required to co- operate with them. Arrangements (which might include 'out of hours' cover) should also be made to enable managers to consult these advisers whenever this is appropriate.

Communication

12   Communication should be a two-way process. Senior management must explain the priority to be given to their aim of restricting exposure so that employees appreciate the need to avoid unnecessary exposure. Managers should also seek feedback about the effectiveness of their policies and about ways of making further improvements.

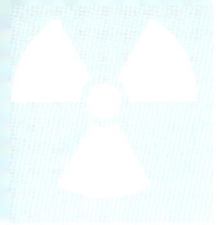

13	The arrangements made in the organisation to restrict exposure should be described in the written local rules, which should be reviewed periodically. By their behaviour senior managers should try to promote a positive culture which emphasises the importance of these arrangements.

14	Employee representatives should be formally consulted about the adequacy of these arrangements through joint health and safety committees or radiation protection committees advised by the RPA. In addition employee representatives should normally be involved in reviews of work practices (see paragraph 57).

15	It may be appropriate in some cases to set up special groups or committees in order to focus attention on the opportunities for further reductions in exposure. Such a group could be established to carry out a special review of work with ionising radiation or to make specific recommendations to management at the development stage of new projects.

Data systems

16	Adequate data systems will be necessary to enable managers and their advisers to make judgements about the restriction of exposure. IRR 85 specifically require certain information to be recorded, for example area and personal monitoring data. Senior management must ensure that this information, together with any job-specific data obtained from special surveys and investigations, is made available to line managers in a suitable form. With such information, and advice from the RPA, managers should be well placed to judge the effectiveness of arrangements to restrict exposure and to identify the need for future action.

Information, instruction and training

17	Employees who work with ionising radiation should be provided with sufficient information, instruction and training to enable them to understand the importance of restricting exposure and the specific working procedures that should be followed. Also they should be advised of their responsibilities under Regulation 6[4] and 34 IRR85. Classified persons and trainees should be given specific information about the medical and technical requirements of IRR 85; they are also more likely to require refresher training.

18   In order to satisfy the requirements of Section 2 of HSWA and Regulation 12 IRR85, training (and selection) of employees should be sufficient to ensure that they are competent to work safely with ionising radiation. In addition the special training needs of individuals or groups of individuals (including young persons and pregnant women) should be assessed. Proposed legislation to implement existing European Directives will also contain requirements for health and safety training[3][4].

19   One important outcome of the training should be to make employees alert to the simple actions they can take to minimise their exposure (and the exposure of others); it should also help them to understand why such actions are worthwhile, even when exposures are relatively small. For example, drivers who deliver packages containing radioactive substances should appreciate the value of keeping their distance from those which have a high transport index; sack trucks, trolleys or nets could be used to transfer the packages to and from the vehicle, rather than carrying them.

20   Employees of outside contractors will also require information (and possibly on-site training). Managers will need to cooperate closely with the contractor, who should receive adequate information about relevant hazards from ionising radiation in the areas of the site where the contractor's employees will work. The information should include a copy of the local rules for the area, containing any written system of work required by Regulation 8(6)b IRR85. They will also need to be satisfied that the contractor's employees have received the training required for work in those areas. (Proposed Regulations to implement an existing European Directive[5] will require the provision of information to contractors.)

21   The better informed employees are about their exposure to ionising radiation the more likely it is that they will modify their work practices to restrict that exposure. For example, if the location of relatively high dose rate zones within 'controlled areas' are clearly marked employees will restrict their exposure by minimising the time spent in such zones. More generally, employees subject to routine dose assessment should have free access to summaries of their own doses as assessed by the employer's dosimetry service.

22   Rehearsal of tasks likely to involve significant exposure to ionising radiation will reduce the time taken to complete the work and hence restrict the exposure resulting from the operation. It could include the use of mock-ups of plant or equipment, computer simulation or the use of video recordings of an actual work practice. Although rehearsal is clearly relevant to planned maintenance in high dose rate areas it may be essential in other situations, for example when making arrangements to deal with foreseeable emergencies such as stuck radiography or radiotherapy sources.

23   More information is given about training for competence and other aspects of organising for health and safety in 'Successful Health and Safety Management'[6].

PRIOR ASSESSMENT OF EXPOSURE

General

24   The duty under Regulation 6 IRR 85 implies that an assessment has to be made of the exposure likely to arise as a result of the work, the associated risks to individuals and any precautions that could be taken to avert these risks (The Management of Health and Safety at Work Regulations 1992[3] contain explicit requirements for risk assessment). A judgement must be made about the need for any additional measures, beyond those already planned, to minimise risks. The time, cost and physical problems of taking the measures should be weighed against the positive benefits associated with any reductions in exposure (including improved productivity, social factors etc). If the 'costs' are so disproportionate to the risk that it would be unreasonable for the employer to incur them, no further action is required. However, the greater the risk the more likely it is that it would be reasonable to go to substantial expense to reduce it.

25   It is assumed that the probability of long term health effects occurring as a result of exposure to relatively low levels of ionising radiations is broadly proportional to the dose of ionising radiation received by the individual. Thus it is usual to take the radiation doses received by employees as an indicator of the risk arising from the work.

26   Often, it is necessary to assess both the doses received by the most exposed individuals and the 'collective dose' - that is the sum of the doses received by all the individuals exposed to ionising radiation as a result of that work. Thus a check can be kept on the overall risk to the workforce. Furthermore, an estimate of 'collective dose' will be necessary for any quantitative assessment of the risk (see paragraph 35).

27   However, paragraph 9 ACoP Part 1[2] explains that, in choosing between restricting collective dose and restricting doses to individuals, priority should be given to keeping individual doses as far below the statutory dose limits set out in IRR85 as reasonably practicable. Furthermore, the effect of Regulation 28 IRR 85, paragraph 172 ACoP Part 1 and of ACoP Part 4[7] should be to encourage the restriction of individual exposure to below an average of 15 mSv per year in all cases.

28   There will be occasions, however, when it is not reasonably practicable to share out the work among a group of employees with particular skills and experience. In these cases it is necessary to achieve a reasonable balance between individual and collective dose, but priority should be given to the restriction of individual exposure as far as possible.

Systematic assessment

29   The advice of the RPA should be sought before carrying out the assessment (paragraph 75 ACoP Part 1) so that all the practicable options to restrict exposure can be examined. The RPA may identify the need to bring in other professionals such as ventilation engineers or control engineers, in appropriate cases.

30   Detailed consideration of the various options may not be necessary because the appropriate measures to restrict exposure are set out in accepted standards or authoritative Codes of Practice (for example references 8 and 9). For many straightforward situations the clear technical advice of the RPA may be sufficient.

31   However, there will be some work activities, particularly those involving new plant or equipment, where it is not clear which protective measures should be

adopted. In these cases a systematic approach is called for to ensure that all the significant factors which distinguish the possible options (for example, dose implications and relative cost of the protective measures) are properly considered.

32   IRR 85 require employers to use physical safeguards to restrict exposure wherever reasonably practicable and, by implication, to set up systems of inspection and preventive maintenance to ensure that these safeguards continue to be effective. (Proposed regulations[4] will contain more explicit requirements). Thus the options of engineering controls, design features and warning devices (and the maintenance of such measures) must be examined first. ACOP Parts 1 and 2 (and, in relation to radon, ACOP Part 3[10]) give practical guidance on the physical safeguards which should be assessed.

33   Attention should then be focused on improvements in systems of work, training etc after the provision of physical safeguards has been fully considered and a judgement has been made about which of these safeguards it would be reasonably practicable to provide.

34   The effort required to make the assessment will depend on the exposure likely to arise from the work activity. In some day to day situations line managers may need to make frequent decisions affecting the exposures of a few individuals. In most cases, it will usually be sufficient for these decisions to be based on the judgement of the RPA (or supporting staff). However, where the assessment concerns a new work activity which has the potential for very significant individual or collective exposure, a much more extensive effort would be required.

Analysis

35   A quantitative technique such as cost benefit analysis, is sometimes seen as a useful way of helping to decide between possible options to restrict exposure. There is no requirement to use such a technique but it may provide valuable information to decision-makers in cases where it is not clear that the 'cost' of adopting a particular measure is grossly disproportionate to the likely benefits arising from lower exposures.

36   However, it must be recognised that such techniques have their limitations and they can only be an aid to the assessment; they cannot, themselves, produce a decision. For example, it may not be possible to quantify all the factors involved such as the balance between collective and individual doses, the rate at which dose is received, and broader social factors; these factors will often need to be addressed separately.

37   If quantitative techniques are used as an aid to assessment, senior management will have to determine the 'value' to be placed on the possible harm arising from the exposure. Usually, this would be done by specifying a monetary value for a unit of collective dose.

38   Detailed guidance on quantitative analysis, including a range of <u>1990</u> values for the unit of collective dose, is given in 'ALARP from theory towards practice[11]'. Although this publication is principally directed to expert advisers some sections may be useful to managers.

DESIGN OF WORK AREAS, EQUIPMENT AND PLANT

39   The greatest opportunities to restrict exposure arise when a new (or substantially modified) work activity is being planned. A prior assessment should enable management to identify design options that eliminate or minimise exposure by engineering methods. It may be possible to use a technique which does not involve radioactive substances or to select highly reliable plant or equipment which will minimise the need for maintenance work in high dose rate (or grossly contaminated) areas or which will enable such work to be done remotely.

40   It is essential that the advice of the RPA should be sought to assist with the prior assessment. Also, it is vital that all those involved in the project, including any outside contractors, appreciate the importance of the ALARP requirement and the design strategy by which this requirement is to be met. The strategy should be referred to in the health and safety policy of the organisation.

41   In appropriate cases senior managers who are responsible for the project could set up special groups or committees to make recommendations about the restriction of exposure. Such groups should ensure that the work of designers, installers, engineers, operational managers and the RPA are brought together.

42 Large organisations may specify dose design criteria for designers, determined by past experience. These criteria can be a useful guide to the designer, but they may not, in themselves, be sufficient to enable the employer to demonstrate that enough has been done to meet the ALARP requirement.

43 Manufacturers and suppliers of equipment for use in work with ionising radiation also have duties in relation to the restriction of exposure under Section 6 of HSWA (as extended by Regulation 32 IRR85); they must ensure that such equipment is designed, constructed and (where necessary) tested so as to restrict exposure so far as reasonably practicable. The supplier should provide prospective users with adequate information about typical dose rates and about any safety features and engineering controls incorporated into the design to restrict exposure. (Regulations to implement the European Machinery Directive[12] will also place duties on manufacturers). The installer of the equipment should undertake a 'critical examination' of the way in which such equipment is installed, in conjunction with a radiation protection adviser (in accordance with Regulation 32 IRR85).

44 The basic design of the installation or work area should be established using both the information provided by suppliers and good engineering practice (based on past experience and current technological developments). Particular attention should be given to the general guidance in the ACOP Parts 1 and 2[2].

45 Once the basic design has been established, an estimate should be made of individual and collective doses likely to result from the routine operation and maintenance of the equipment or plant in the work area. In particular the major contributions to these doses should be identified.

46 If any feature of the design is identified as being likely to give rise to significant individual or collective dose this should be considered systematically. A reasonable attempt should then be made to 'quantify' the judgement that the chosen design option is consistent with the ALARP requirement. Figure 1 illustrates the type of procedure that might be used.

47 A record should be made of any significant findings and assumptions together with any estimated doses associated with the chosen option; these will assist

the employer in any later review of the work. Employees' representatives should normally be advised of these findings through the joint health and safety committee, radiation protection committee or other relevant group.

REVIEWS OF WORK WITH IONISING RADIATION

Need

48   Managers and their advisers can make ad hoc changes to current work practices (including routine maintenance) to further restrict exposure, but the most effective means of ensuring that the ALARP requirement is being met is to carry out a formal review.

49   A special review of the work may be shown to be necessary either by a health and safety audit or by adverse dose trends which reveal significant departures from (or inadequacies of) local rules. In addition, employers should normally review their work with ionising radiations, periodically, because:

(a)   adequate attention may not have been given to the ALARP requirement when plant or equipment was first installed or when local rules were prepared;

(b)   workloads may have increased or work practices may have altered as a result of staff changes or the introduction of new equipment;

(c)   new procedures may have become available which replace techniques based on the use of ionising radiation;

(d)   improved control techniques and standards may have been developed;

(e)   safety features and warning devices may not have been adequately maintained;

(f)   a review may enable managers to identify particular work practices (or exposure routes) which are leading to significant and unnecessary exposure of individuals or groups of individuals.

50   In addition, Part 4 of the Approved Code of Practice (ACOP)[7] has drawn attention to increases in the estimates of risk associated with exposure to ionising radiation. In consequence, earlier decisions about the precautions necessary to satisfy the ALARP requirement may need to be re-examined.

51   Furthermore, when certain designated dose levels are exceeded the employer should arrange for investigations to be carried out. These action levels are a means of warning managers that exposures may not be as low as reasonably practicable. The relevant levels are individual recorded doses which:

(a)   exceed 15 mSv within a calendar year (Regulation 28 IRR 85) or

(b)   reach a cumulative dose of 75 mSv or more within any five consecutive calendar years starting from 1.1.88 (ACOP Part 4);

(c)   reach any additional investigation levels (below these figures) set by senior management to ensure that the ALARP requirement is being met.

52   A further review may be required when a recorded dose exceeds either a relevant dose limit (Regulation 29 IRR85) or 30 mSv within a calendar quarter (implied by Regulation 13(3)(f) IRR85). Senior management should also consider setting an investigation level for collective (or average individual) dose for the workforce (as is now done by provision of a specific licence condition in relation to licensed nuclear sites).

Conduct of reviews

53   If it is believed that individual and collective doses are low, a periodic review will still be necessary to confirm that this is the case. However, the effort required to carry out the review will be small, commensurate with the potential to make further reductions in doses.

54   Periodic reviews or investigations should normally involve a critical and systematic assessment of the current controls and work procedures. This assessment should take account of current dose levels and of any estimates of

future trends in these levels (deteriorating plant or equipment may lead to increased maintenance and thus to greater exposures). It should identify options for further restricting exposure.

55 The review process should be transparent so that it will be clear to the workforce and their representatives how decisions were reached about the need for further precautions (if any). Consequently, it is more likely that these decisions will be accepted by those who work with ionising radiation and additional precautions will be implemented effectively.

56 Senior management should have designated managers at an appropriate level in the organisation (eg line management) to conduct the reviews and to decide what action (if any) needs to be taken as a consequence. The advice of the RPA should be sought for a consistent approach to reviews. However, final responsibility for the outcome of the review will rest with the designated manager, except where the levels of recommended expenditure are such that the review has to be referred to senior management for approval.

57 Figure 2 gives an example of a systematic review to focus attention on those aspects of the work which offer most scope for further reductions in exposure. It should be clear from health and safety audit reports, from job-specific dose information, or from observation, which aspects need to be considered. The review should include information about current work procedures obtained from workplace inspections and from interviews with staff and should take account of doses incurred during inspection, measurement and routine maintenance. Line managers and safety representatives should always be consulted when a review is being undertaken. Appendix 1 shows typical sources of information for employee safety representatives and others involved in any review.

58 The scope for improving shielding or other physical safeguards to the standards of modern plant and equipment may be limited; however some improvements in shielding, ventilation or layout may be reasonably practicable, particularly if improved productivity, greater reliability and other such factors are taken into account. Some organisations have achieved considerable success in restricting exposure by:

(a)   providing additional shielding to reduce the number and size of zones within 'controlled areas' subject to relatively high dose rates;

(b)   re-routing means of access to work areas to avoid relatively high dose rate zones;

(c)   removing radioactive sources from work areas or reducing their activity, for example radioactive deposits in pipework, or timing operations to take maximum advantage of radioactive decay before starting maintenance work; and

(d)   using robotics or other means of remote handling.

59   Appendix 2 contains a checklist for employee safety representatives and others taking part in reviews. It shows a range of precautions that could be considered during a review.

60   The main purpose of the systematic review is to enable managers to take informed decisions, based on the best available information, and to ensure that proper consideration is given to the options for reducing exposure. In most cases it will be clear that particular operations are giving rise to the majority of any individual or collective dose. Occasionally, quantitative analysis may help in deciding which protective measures are worthwhile.

Records of reviews

61   It is important to record the main points of these reviews including:

(a)   any decisions about the adequacy of current controls and procedures;

(b)   details of any action programme (or recommendations to senior management);

(c)   the weight given to any key factors which influenced the decision; and

(d)   the justification, in broad terms, for rejecting any options.

62   Where the review is part of a special investigation concerning an over-exposure (paragraph 52) an indication should be given of the point at which it is considered that any further expenditure on protective measures would be in gross disproportion to the risks that could be avoided by implementing those measures (see paragraph 24).

63   Serious weaknesses in the availability of job-specific information which hampered the review should also be identified so that action can be taken to collect such data for use in any future reviews.

PRIOR ASSESSMENT FOR SPECIAL PROJECTS

64   Routine inspection and maintenance tasks should be examined as part of any periodic review. However, any major maintenance or decommissioning project, which may have significant implications for the exposure of individuals to ionising radiation, calls for a different approach. In such cases the performance of the work may need to be reviewed several times during the course of the project. Examples would include maintenance work carried out in high dose rate areas during power reactor shutdowns and refurbishment of grossly contaminated work areas.

65   These operations need very careful planning. A separate prior assessment is justified because of the doses that could be incurred. The precautions to be considered are likely to include the use of special tools for remote inspection and handling (and other distancing techniques), use of portable shielding, personal protective equipment and rehearsal (using mockups, computer simulation or video recordings of the work area) to limit the time spent in high dose rate (or contaminated) areas. Controlled access (perhaps by permit to work system) will be necessary.

66   The prior assessment should lead to estimates of the expected individual and collective doses for the project. Daily or weekly dose reference levels should be specified on the basis of these estimates in order to trigger a prompt review of the project if actual rates of exposure are significantly greater than predicted. The review might identify the need for better training of a particular individual or for a change to the chosen method of work.

67   An example of the procedure that might be used for assessing and reviewing special projects is outlined in Figure 3.

**Figure 1   OUTLINE EXAMPLE OF PRIOR ASSESSMENT
FOR NEW WORK ACTIVITY**

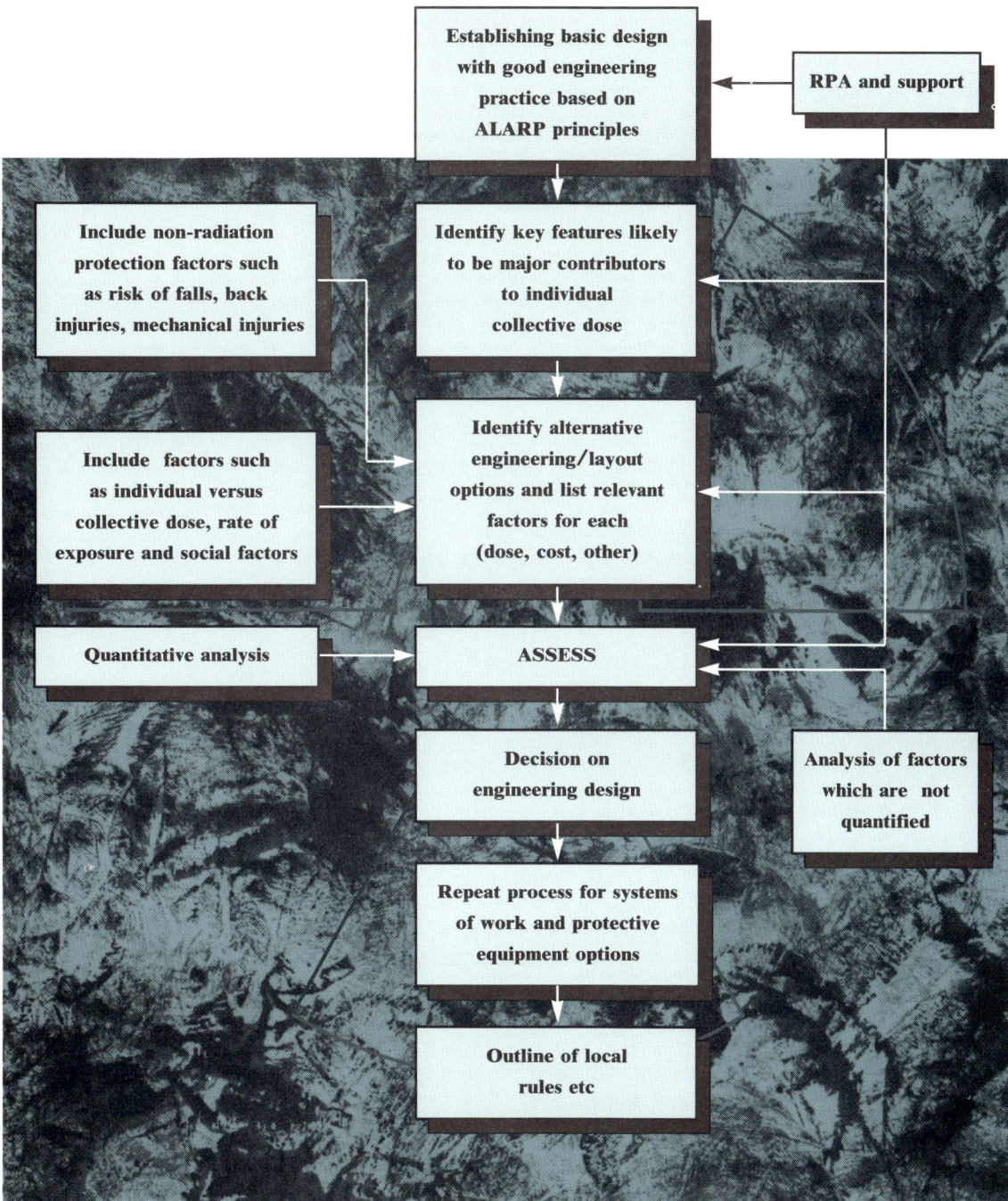

Establishing basic design with good engineering practice based on ALARP principles

RPA and support

Include non-radiation protection factors such as risk of falls, back injuries, mechanical injuries

Identify key features likely to be major contributors to individual collective dose

Include factors such as individual versus collective dose, rate of exposure and social factors

Identify alternative engineering/layout options and list relevant factors for each (dose, cost, other)

Quantitative analysis

ASSESS

Decision on engineering design

Analysis of factors which are not quantified

Repeat process for systems of work and protective equipment options

Outline of local rules etc

Figure 2   **OUTLINE EXAMPLE OF FORMAL REVIEW OF**
OCCUPATIONAL EXPOSURE LEVELS

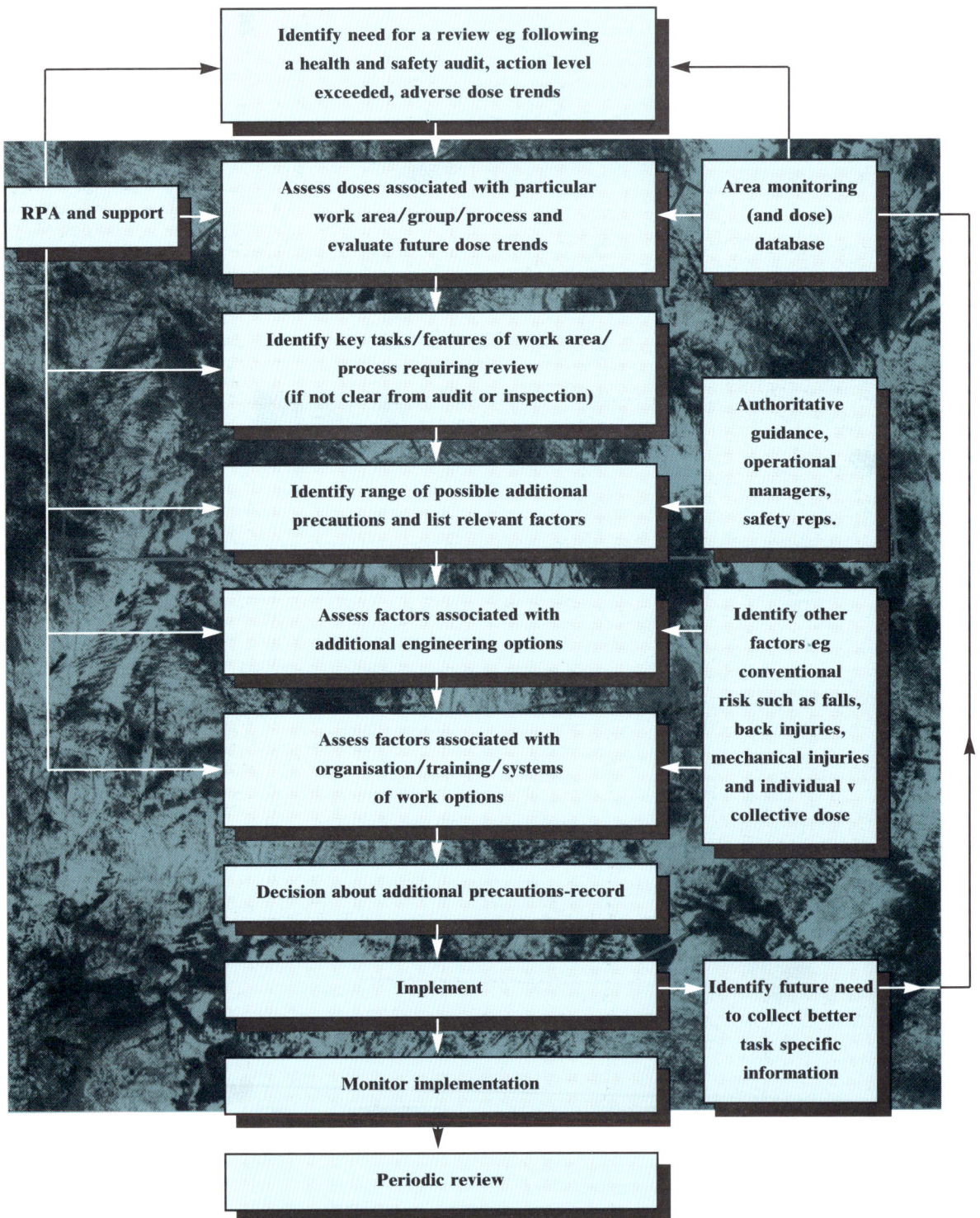

Identify need for a review eg following a health and safety audit, action level exceeded, adverse dose trends

RPA and support

Assess doses associated with particular work area/group/process and evaluate future dose trends

Area monitoring (and dose) database

Identify key tasks/features of work area/ process requiring review (if not clear from audit or inspection)

Identify range of possible additional precautions and list relevant factors

Authoritative guidance, operational managers, safety reps.

Assess factors associated with additional engineering options

Identify other factors eg conventional risk such as falls, back injuries, mechanical injuries and individual v collective dose

Assess factors associated with organisation/training/systems of work options

Decision about additional precautions-record

Implement

Identify future need to collect better task specific information

Monitor implementation

Periodic review

Figure 3   OUTLINE OF SYSTEMATIC ASSESSMENT FOR SPECIAL PROJECTS

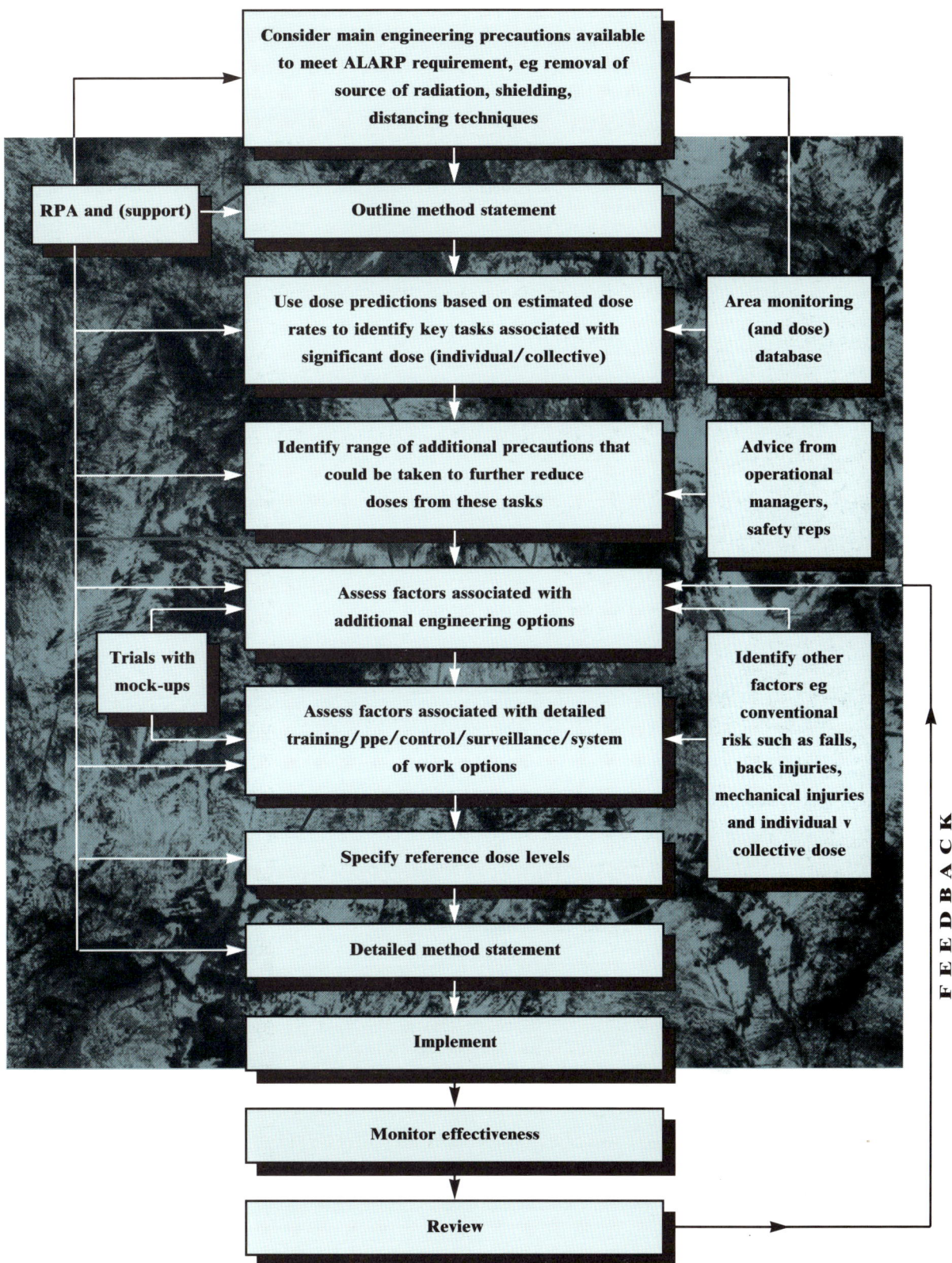

```
                    ┌─────────────────────────────────────┐
                    │ Consider main engineering precautions │
                    │ available to meet ALARP requirement,  │
                    │ eg removal of source of radiation,    │
                    │ shielding, distancing techniques      │
                    └─────────────────────────────────────┘
                                    │
   ┌──────────────────┐   ┌─────────────────────────────────┐
   │ RPA and (support)│──▶│      Outline method statement    │
   └──────────────────┘   └─────────────────────────────────┘
                                    │
        ┌─────────────────────────────────────┐   ┌──────────────┐
        │ Use dose predictions based on        │   │ Area         │
        │ estimated dose rates to identify key │◀──│ monitoring   │
        │ tasks associated with significant    │   │ (and dose)   │
        │ dose (individual/collective)         │   │ database     │
        └─────────────────────────────────────┘   └──────────────┘
                                    │
        ┌─────────────────────────────────────┐   ┌──────────────┐
        │ Identify range of additional         │   │ Advice from  │
        │ precautions that could be taken to   │◀──│ operational  │
        │ further reduce doses from these tasks│   │ managers,    │
        └─────────────────────────────────────┘   │ safety reps  │
                                    │              └──────────────┘
        ┌─────────────────────────────────────┐
        │ Assess factors associated with       │◀──
        │ additional engineering options       │
        └─────────────────────────────────────┘   ┌──────────────┐
   ┌──────────────┐            │                   │ Identify     │
   │ Trials with  │   ┌─────────────────────────┐  │ other factors│
   │ mock-ups     │──▶│ Assess factors associated│  │ eg           │
   └──────────────┘   │ with detailed training/  │◀─│ conventional │
                      │ ppe/control/surveillance/│  │ risk such as │
                      │ system of work options   │  │ falls, back  │
                      └─────────────────────────┘   │ injuries,    │
                                    │               │ mechanical   │
        ┌─────────────────────────────────────┐    │ injuries and │
        │       Specify reference dose levels  │    │ individual v │
        └─────────────────────────────────────┘    │ collective   │
                                    │               │ dose         │
        ┌─────────────────────────────────────┐    └──────────────┘
        │       Detailed method statement      │
        └─────────────────────────────────────┘
                                    │
        ┌─────────────────────────────────────┐
        │               Implement              │
        └─────────────────────────────────────┘
                                    │
        ┌─────────────────────────────────────┐
        │          Monitor effectiveness       │
        └─────────────────────────────────────┘
                                    │
        ┌─────────────────────────────────────┐
        │                Review                │
        └─────────────────────────────────────┘
```

FEEDBACK

APPENDIX 1

TYPICAL CATEGORIES OF SOURCES OF INFORMATION FOR SAFETY REPRESENTATIVES AND OTHERS INVOLVED IN REVIEWS

(a) Details of the work area, plant and equipment including dose rates and typical contamination levels.

(b) Information from operational management, operators and maintenance personnel about current patterns of work, including occupancy times.

(c) Dose information - distribution of individual doses among employees in that part of the site, distribution with time (trends) and, where appropriate, collective dose, task specific doses and the number of individuals close to any investigation levels.

(d) Structure of management.

(e) Levels of training of personnel set against competencies required for that group of individuals.

(f) Relevant written instructions including local rules.

(g) Relevant guidance, for example ACOP[2] Medical and Dental Guidance Notes[8] Radiation Safety for site radiography[9] and publications produced by organisations such as the National Radiological Protection Board, Institute of Physical Sciences in Medicine etc.

(h) Radiation protection adviser, safety adviser, health physicist etc.

APPENDIX 2

RANGE OF PRECAUTIONS THAT MAY NEED TO BE CONSIDERED DURING REVIEWS

(a)   Replacement of equipment, use of new technology, upgrading of engineering controls such as shielding, containment of unsealed sources and ventilation to reduce exposure in working areas.

(b)   Removal or reduction in activity of radiation sources eg by draining tanks or pipework or by allowing for radioactive decay before starting work.

(c)   Improvements in instrumentation, detectors, warning devices.

(d)   Removing items for maintenance from high dose rate areas to low dose rate areas.

(e)   A reduction in the need for manual intervention.

(f)   Distancing techniques including remote handling, inspection and measurement.

(g)   Restricting frequency of maintenance necessary in high dose rate areas (eg re-scheduling work) and need for inspection and measurement.

(h)   Demarcation of high dose rate areas, dose contour maps etc and designation of access routes and rest facilities outside high dose rate areas.

(i)   Minimising time spent in high dose rate areas by task redesign, rehearsal and practice on mock-ups.

(j)   Authorised entry procedure, permits to work (some routine maintenance operations).

(k)   Improved training and provision of information for managers and persons working with ionising radiation (including contractors) about risks and basic precautions.

(l)   Improvements in management organisation, defining responsibilities.

(m)   Improvements in supervision, task management.

(n)   Improvements in task specific dose information.

REFERENCES

1   The Ionising Radiations Regulations 1985 SI 1985/1333
     HMSO 1985 ISBN 011 057333 1

2   HSC *Protection of persons against ionising radiation arising from any work activity: the Ionising Radiations Regulations 1985: Approved Code of Practice* COP 16
     HMSO 1985 ISBN 011 883838 5

3   The Management of Health and Safety at Work Regulations 1992 SI 1992/2051
     HMSO 1992 ISBN 011 025051 6

4   European Communities Council *Directive of 30 November 1989 concerning the minimum safety and health requirements for the use of work equipment by workers at work* (89/655/EEC) 30 December 1989, vol 32, no L393,13-17

5   European Communities Council *Directive of 4 December 1990 on the operational protection of outside workers exposed to the risk of ionising radiation during their activities in controlled areas* (90/641/EURATOM) 13 December 1990, vol 33, no L349, 21-25

6   HSE *Successful health and safety management* HS(G)65 HMSO 1991 ISBN 011 885988 9

7   HSC *Dose limitation - restriction of exposure: Additional guidance on regulation 6 of the Ionising Radiation Regulations 1985 Approved Code of Practice Part 4* L7
     HMSO 1991 ISBN 011 885605 7

8   HSE National Radiological Protection Board *Guidance Notes for the protection of persons against ionising radiations arising from medical and dental use*
     HMSO 1988 ISBN 0 85951 2991

9   Oil and Chemical Plant Constructors' Association *Radiation safety for site radiography* 1986 Brentford, Kluwer Publishing

10   HSC *Exposure to radon: The Ionising Radiations Regulations 1985 Approved Code of Practice Part 3* COP 23 HMSO 1988 ISBN 011 883978 0

11   Stokell, P J, Croft, J R and others *ALARA from theory towards practice* A joint report to the Commission of the European Communities by the National Radiological Protection Board and the Centre d'Etude sur l'Evaluation de la Protection dans le Domaine Nucleaire (EUR 13796) Radiation protection series
     HMSO 1992 ISBN 92 826 3274 1

12   European Communities Council *Directive of 14 June 1989 on the approximation of the laws of the Member States relating to machinery* (89/392/EEC) 29 June 1989, vol 32, no L183, 9-32 as amended by Directive 91/368/EEC.

Printed in the United Kingdom for HSE, published by HMSO   C50   11/92